571.8

LIFE'S CYCLES

FROM
BIRTH to DEATH

IRENE YATES

ILLUSTRATED BY
GRAHAM AUSTIN

The Millbrook Press
Brookfield, Connecticut

Published in the United States
in 1997 by

The Millbrook Press, Inc.
2 Old New Milford Road
Brookfield, Connecticut 06804

First published in Great Britain
in 1996 by

Belitha Press Limited
London House
Great Eastern Wharf
Parkgate Road
London SW11 4NQ

Text © 1996
Belitha Press Limited
Illustrations © 1996
Graham Austin

Editor: Maria O'Neill
Designer: Frances McKay
Consultant: Steve Pollock

All rights reserved
Printed in China

Library of Congress Cataloging-in-Publication Data

Yates, Irene.
    From birth to death / Irene Yates ; Illustrated by Graham Austin.
      p. cm. — (Life's cycles)
  Includes index.
  Summary: Follows a year in the life of the animals and plants occupying a pond environment, showing the life cycles and interactions of the various species.
   ISBN 0-7613-0303-0 (lib. bdg.)
   1. Life cycles (Biology)—Juvenile literature.   2. Pond ecology—Juvenile literature.   [1. Pond ecology. 2. Ecology.] I. Austin, Graham, ill. II. Title.   III. Series.
QH501.Y38   1997
571.8—dc21   97-13057
                                                                CIP
                                                                 AC

CONTENTS

Life Cycles in a Pond 4
The Pond 6
Winter Quiet 8
A New Beginning 10
The Pond Comes to Life 12
Summer Begins 14
High Summer 16
The End of Summer 18
Autumn Begins 20
Preparing for Winter 22
Winter Again 24
All Year Round 26
Glossary 28
Index 30

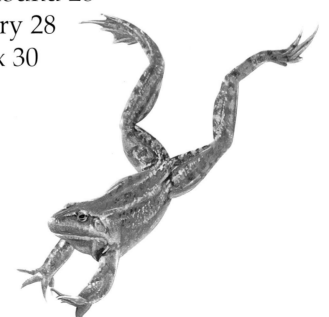

LIFE CYCLES IN A POND

A pond is a whole world of living creatures and plants. The lives and **life cycles** of the different creatures and plants are all linked. Each one has its own place in the **food chain**. The birth, life, and death of all the plants and animals are carefully balanced so that the pond and its wildlife survive. If one plant or animal disappears, the pond's wildlife is put in danger.

THE LIFE CYCLE OF A FROG

The male frog clings to the female frog's back during mating. The female can lay up to 3,000 eggs at one time.

The female frog lays eggs. They sink to the bottom of the pond and then float to the surface.

Tadpoles hatch from the eggs in late May. Many of them are eaten by other animals.

The hungry pike waits for the tadpoles to swim by. Pikes have large mouths with backward sloping teeth for catching their prey.

As the seasons of the year go by, the wildlife in the pond changes. Animals are born, seedlings grow into plants, fish **spawn**, and the eggs of insects, birds, and **amphibians** hatch. As summer comes, if the young animals are not eaten, they begin to grow into adults.

If they are not eaten, the young frogs spend the summer sitting near the pond bank catching flies in the sun.

The heron stands in the water and waits patiently. When he spies some fish or tadpoles, he spears them with his bill.

The tadpoles grow back legs first, then front legs. They begin to look like tiny froglets. The froglets slowly lose their tails and become young frogs.

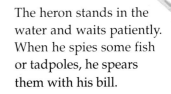

Words in **bold** are explained in the glossary on pages 28-29.

THE POND

Welcome to the pond. Many plants grow around the water's edge. They have strong stems that creep through the mud at the bottom of the pond. Many of the pond animals feed on the plants and shelter among the leaves.

How many animals and plants can you find in the picture? You'll see them many times throughout this book as we follow their lives. Read the story and look at the pictures to learn about the creatures and plants that live around the pond. Here the pond is shown in high summer, but it changes as the year goes by.

WINTER QUIET

It is the beginning of the year. Winter is coming to an end. Most of the plants have died back. The pond is very quiet. Sometimes a bubble breaks the surface of the water where there is a fish below. Most of the animals are **hibernating** or are in the resting stages of their life cycles. The pond freezes over, but the creatures underneath can survive if some of the water stays unfrozen.

◀ Water voles swim around the pond looking for food. They feed on water plants, insects, and snails.

▲ Some **species** of geese fly from the far north to spend the winter near more southern ponds, where the weather is warmer.

▲ Water voles do not hibernate in winter. They live on land in tunnels called burrows. The female makes a nest in a bank or in a hollow tree in February, ready for mating.

▼ The coot is a bird that dives underwater to collect plants and insect **larvae** to eat.

▲ Sticklebacks like the shallow water at the edge of the pond. They feed on worms, snails, and small **crustaceans**.

▲ Dead leaves and plants form a rich layer at the bottom of the pond. Sometimes frogs bury themselves in it to hibernate. Insect larvae feed on the layer all winter.

A NEW BEGINNING

As spring arrives, the pond world begins to wake up. The water plants sprout new leaves and seeds begin to **germinate**. Plants at the edge of the pond give shelter to the water birds which use their leaves and twigs to build nests. All the toads and some frogs spent the winter hibernating on land. In spring, they return to the pond to **breed** and have their young.

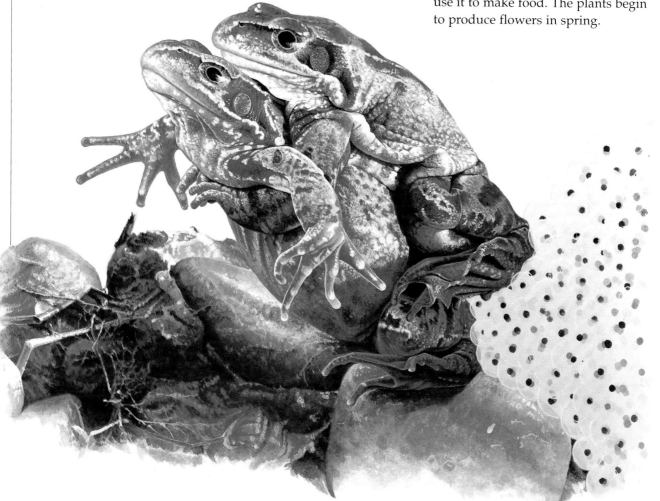

◀ The male frog clings to the female during mating. She lays her **spawn** in the pond. The spawn floats to the surface. A frog can lay more than 3,000 eggs at one time and a toad can lay 6,000 eggs at once.

▲ Each plant has roots, a stem, and leaves. The leaves take in sunlight and use it to make food. The plants begin to produce flowers in spring.

▶ Mallards build shallow nests on the ground. They lay eggs from March to May. The mother's body and soft **down** keep the eggs warm. Her feathers are drab so that she cannot be seen in the reeds.

THE POND COMES TO LIFE

Spring has arrived. Animals are getting ready to have their young. Birds are building their nests and looking for a mate. The fish are beginning to spawn. Insect eggs are starting to hatch as the weather becomes warmer. Some birds return to the pond after spending the winter in warmer places.

▼ Pike spawn in the shallows. Large females swim along with two or three smaller males. They lay their eggs over underwater plants and the males **fertilize** them. The adults swim off and leave the eggs in the water.

▼ A pair of grebes meet. They swim and dive together and catch weed in their beaks in a **courtship dance**. After this dance, the grebes mate.

▶ Tadpoles hatch from the frog eggs. They start to eat the plants at the bottom of the pond. Their back legs are starting to grow.

SUMMER BEGINS

The plants begin to produce bright flowers which **attract** insects. The insects eat **nectar** and help **pollinate** the flowers. There is new life everywhere. The waterbirds' eggs hatch. Young fish hatch from their eggs to swim freely.

▲ The female mosquito lays between 150 and 350 eggs. The eggs float like a tiny egg raft. After a few days they hatch into larvae which hang down for three to five days. Many creatures feed on them.

◀ Newts attach single eggs to plants in early spring. In early summer the eggs hatch into newt tadpoles. The parents stay in the pond until the end of the summer.

▶ The female stickleback is fat with eggs, which will soon hatch. The male brings her to the nest where she will lay the eggs.

▼ The duck's eggs have hatched. When the ducklings' down is dry and fluffy, the mother leads them to the water. They can swim and run and feed themselves right away.

► The yellow flag iris has two to three flowers. Water snails and small crustaceans live on it.

► Frog tadpoles grow back legs first, then front legs. The tail slowly disappears and the tadpole looks like a frog.

▲ The dragonfly eggs that were laid last summer have hatched into nymphs. Nymphs eat water fleas, tadpoles, and small fish.

HIGH SUMMER

Summer is here and the pond is full of life. There is plenty of food for birds, bats, and dragonflies as swarms of insects hover over the pond. The weather becomes warm and mosquitoes and midges begin to hatch.

▼ Swallows catch their food in flight. There are many different kinds of insects around the pond for the swallows to feed on.

▼ Young moorhens can swim within hours of hatching, and they follow their mother around the pond.

▶ In mid July, small froglets jump out on to the pond bank. They will grow into full-sized frogs over the next three years.

◄ The swallowtail butterfly hovers above the pond. Some swallowtails live close to wetlands and rivers. Others live in fields and meadows.

◄ The nymph dragonfly leaves behind a dry case when it becomes an adult. Dragonflies have **compound eyes** which help them to catch insects and flies in flight.

◄ Young grebes and dabchicks are more timid than young moorhens. They hide at the first sign of danger.

► The young voles are ready to leave the nest and make burrows of their own.

THE END OF SUMMER

It is late summer. The pond has become overgrown. The birds, animals, and fish have finished breeding. There is plenty of food for all the wildlife that lives in and around the pond. Some of the animals and birds are fattening themselves up for winter.

▼ Grass snakes wait near the pond to feel the **vibrations** of animals that approach. Then they pounce. Grass snakes are good swimmers and feed on fish, frogs, and toads. They swallow their prey alive.

▼ Water lilies float on the surface of the pond but are anchored to the bottom. The petals close at night and open in the morning for pollination.

◄ Pike hide among the reeds at the edge of the pond where they cannot be seen. They ambush fish or any other creature that comes by.

◄ Late summer is a good time for **anglers.** The fish snatch at the angler's bait because they are trying to eat enough to survive the long winter ahead.

► A heron flies in for a visit. His long legs wade through the water and he stands very still. He is watching the creatures in the pond. When anything swims by, he grabs it with his bill.

► Newts leave the pond to live on land. They will return to the water to breed in two to three years. Newts are not often eaten by other creatures because they don't taste very nice.

AUTUMN BEGINS

As autumn begins, there are still sunny days when adult insects skim the surface of the pond. The plants **disperse** their seeds. Some seeds are in berries which animals eat and spread as droppings. Other seeds become trapped in the fur of animals that brush by them, such as the otter.

▼ The dragonfly has laid its eggs. It will soon die, as the colder weather approaches.

◄ The water vole searches for grass and reeds at the water's edge. If danger threatens, it scurries back to its burrow or dives into the water.

► In autumn, diving beetles eat anything they can find, including small fish and newts. They visit the surface regularly to breathe.

► The pike lurks at the bottom of the pond. Pike have excellent eyesight for spotting their prey.

▲ The swallows fly low over the pond surface, feeding on insects before **migrating** south for winter.

▼ At dusk, the otter comes to the pond. Its long sleek body and webbed feet help it to swim fast after fish.

▼ The young ducks are now almost fully grown. They eat as much as they can before winter sets in and food becomes scarce.

PREPARING FOR WINTER

It is autumn and many of the plants die. The leaves turn brown, **wither**, and fall to the bottom of the pond. Some plants are still in flower. They produce seeds and fruit which small animals collect for their winter supplies. Frogs, toads, and newts search for good places to hibernate. The adult insects die. Their eggs and larvae survive the winter to become adults next year.

▼ Water voles don't hibernate. They come out of their burrows to feed all year round. In autumn they store grass stems in their burrows for winter when they won't be able to find any fresh food.

► Deer come to drink at the water's edge. The male deer will soon shed its antlers only to grow another pair next year. Deer spend the winter months living on roots and berries.

◄ Winter visitors have arrived. New birds come to the pond from colder places to spend the winter months here. These golden eye ducks will disappear again next spring when the warmer weather arrives.

► Older frogs burrow into the soft mud at the bottom of the pond. Young frogs, toads, and newts usually hibernate on land in holes in the ground.

WINTER AGAIN

Winter comes to the pond. The weather turns cold and the air quickly becomes cooler. Ice forms on the surface of the pond which helps insulate the pond water underneath.
Light passes through the ice on the surface of the pond and helps the plants in the pond to survive.

▼ Swans spend all year on or around the pond. Some swans migrate from cold to warmer areas to spend the winter in a milder place.

▼ The otter dives under the ice looking for fish to feed on. Bubbles of air are caught in its thick furry coat.

▲ Herons live near the pond all year round. They hunt for fish and frogs.

► Cold weather and frost kills the pond plants and they die back in winter.

▼ Fish can live under the ice, but they move around less than during the rest of the year.

▼ Caddis fly larvae make cases of sand, twigs, and leaf fragments to protect themselves. The larva has hooks on its body which grip the inside of the case.

ALL YEAR ROUND

A whole year has passed and the pond and its wildlife have come full circle through four seasons. The cycle begins again in spring. Here you can see the life cycle of an insect, the dragonfly. On the opposite page you can see the life cycle of a bird, the mallard duck.

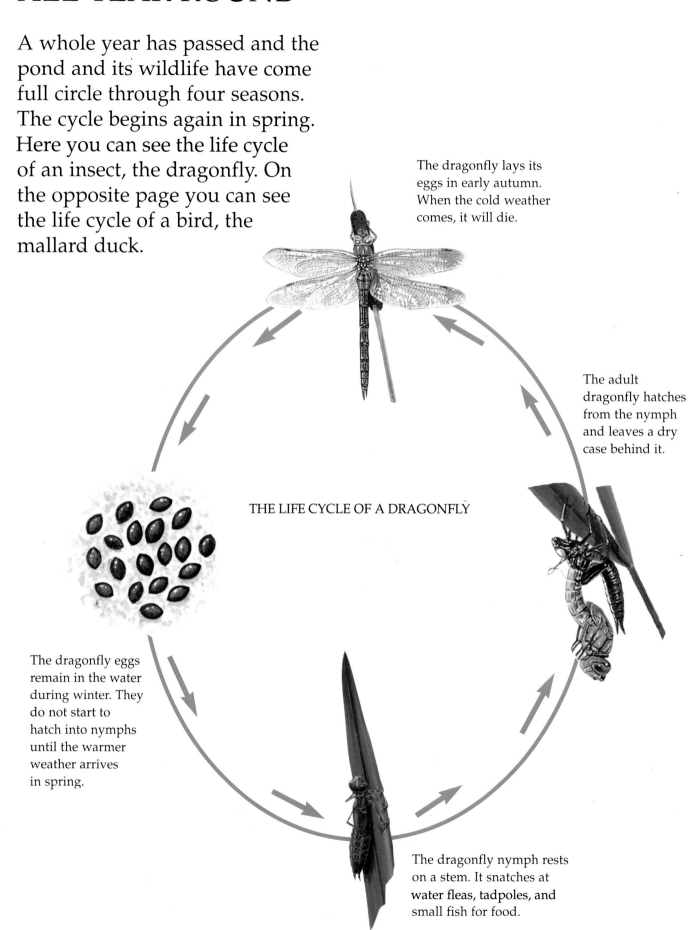

The dragonfly lays its eggs in early autumn. When the cold weather comes, it will die.

The adult dragonfly hatches from the nymph and leaves a dry case behind it.

THE LIFE CYCLE OF A DRAGONFLY

The dragonfly eggs remain in the water during winter. They do not start to hatch into nymphs until the warmer weather arrives in spring.

The dragonfly nymph rests on a stem. It snatches at water fleas, tadpoles, and small fish for food.

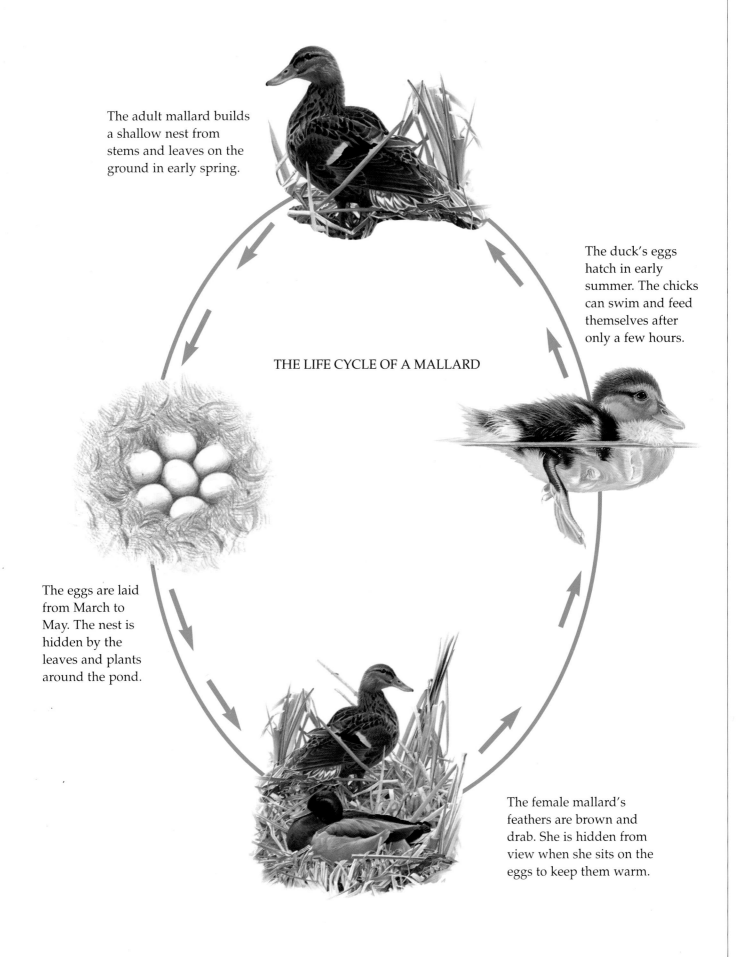

GLOSSARY

amphibian A group of animals, such as newts or frogs, which spend part of their life in water and part on land.

anglers People who catch fish using fishing rods. Angling is a hobby.

attract To draw or pull towards something.

breed To produce young.

compound eyes Eyes which are made up of lots of parts. Compound eyes help insects to see in all directions at once.

courtship dance This happens when two animals, such as grebes, meet and attract each other. They carry out this to show they like each other and want to mate.

crustacean An animal with a hard shell on the outside of its body, such as a crab.

disperse To scatter or spread out.

down The soft, fluffy feathers on a duck or young bird.

fertilize To join a special male cell such as sperm to a special female cell (the egg) to produce young. For example, the male pike's sperm has to join to the female's egg to fertilize it before it can grow into a young pike.

food chain A series of living things that depend on each other for food. For example, a rabbit eats grass and a fox eats rabbits. This is a food chain.

germinate To start to grow. In plants, seeds and spores germinate and grow into new plants.

hibernate, hibernating To spend the winter asleep.

larva, larvae The soft wormlike creature which hatches from an insect's egg is called a larva. More than one of these creatures are called larvae. The caterpillars of butterflies are larvae.

life cycles The changes which an animal passes through during its life. These begin when an egg is fertilized and grows into an adult. The last stage of an animal's life cycle is when it dies.

lungs The parts of the body which an animal uses to breathe. Our lungs are found in our chest.

mate, mating To meet another creature of the same **species** to breed and produce young.

migrate, migrating To move from one area to another every year. Birds such as swallows migrate to warmer areas during winter.

nectar A sugary liquid produced by flowers which insects like to eat. Bees use nectar to make honey.

nymph A young insect, such as a dragonfly, before it grows wings and becomes an adult.

oxygen A gas which is part of air. We cannot see air or oxygen. Animals and plants need oxygen to live. Some animals breathe air into their lungs to get oxygen into their bodies.

pollinate To carry pollen from the male part of a flower to the female part of a flower. The pollen may be carried by the wind, or by insects as they fly from flower to flower, feeding on nectar.

spawn The eggs of fish, frogs, toads, and other amphibians. Spawn is laid in water and surrounded by clear jelly. Also, the act of laying the eggs.

species One particular type of animal or plant. For example, a moorhen is a species of bird and a stickleback is a species of fish. One species cannot breed (have young) with another species. A moorhen can only breed with another moorhen and not with a grebe or any other bird.

vibrations Shaking or trembling movements which are made when something passes by. The snake feels the earth move or vibrate when a creature comes close.

wither To dry up and die.

INDEX

Words in **bold** appear in the glossary on pages 28 and 29

angler 19
autumn 20-21, 22-23

bat 16
berries 20, 23
burrow 9, 17, 20, 22

compound eyes 17
coot 9
courtship dance 13
crustacean 9, 15

dabchick 17
deer 23
diving beetle 20
down 11, 15
dragonfly 15, 16, 17, 20, 26
duckling 15, 21

egg 4, 10, 11, 12, 13, 14, 15, 20, 22, 26, 27

feathers 11, 27
fertilization 12
food chain 4
frog 4-5, 9, 10, 15, 16, 18, 22, 23, 25
frost 25
fruit 22

geese 9
germination 10
golden eye duck 23
grass snake 18
grebe 13, 17

heron 5, 19, 25
hibernation 8, 22, 23

ice 8, 24, 25

insect 8, 14, 16, 17, 20, 21, 22

larvae 9, 14, 22, 25

mallard 11, 27
mating 4, 10, 13
midge 16
migration 21, 24
moorhen 16
mosquito 14, 16

nectar 14
nest 9, 10, 11, 12, 14, 17, 27
newt 9, 14, 19, 20, 22, 23

otter 20, 21, 24

pike 4, 12, 18, 20
pollination 14, 18

reeds 11, 18, 20
roots 10, 23

seed 10, 20, 22
snail 8, 9
spring 10-11, 12-13
stickleback 9, 14
summer 6-7, 14-15, 16-17, 18-19
swallow 16, 21
swallowtail butterfly 17
swan 24

tadpole 4, 5, 13, 14, 15
toad 10, 18, 22, 23

water lily 18
water snail 15
water vole 8, 9, 17, 20, 22
winter 8-9, 18, 21, 22-23, 24-25
worm 9

yellow flag iris 15